A Glimpse of DARTMOOR PLACE NAMES

Andrew Stevens

Peninsula Press

The map of Dartmoor on pages 16-17 is reproduced with the kind permission of the Dartmoor National Park Authority.

Line drawings by Sue Capey.
Other illustrations by Brian Ainsworth.

Published by Peninsula Press Ltd
P.O. Box 31
Newton Abbot
Devon TQ12 5XH

Tel: 0803 875875

Printed in England by D.D.S. Colour Printers, Weston-Super-Mare.

ISBN 1 872640 09 5

A GLIMPSE OF DARTMOOR · PLACE NAMES ·

Contents

Introduction

Dartmoor was given National Park status in 1951. This area included the ancient Forest of Dartmoor and the surrounding border parishes with their common lands. The title 'Forest' therefore, is not a reference to trees, because, although more wooded than at present, (discounting Forestry Commission conifer plantations), it was never a true forest. It is called Forest because of its status from the 13th century as a Royal hunting ground. Strictly speaking, when King Henry III gave the area to his brother Richard in 1239, it automatically became a 'chase' technically, because a Forest could only be possessed by the sovereign. Richard was then only the Duke of Cornwall, and since that time, Dartmoor has largely remained Duchy property - even if the title of Forest has persisted.

The place names of Dartmoor are mainly of Saxon origin because, in most areas, there was some time between the withdrawal of the Celts and takeover by the Saxons. As a result, only some of the Celtic names survived. The names of many physical locations are relatively recent. Virtually every landmark, however small, has a name as a reference point for knowing where you are and for giving directions to someone else.

Obviously there are altogether too many place names to explain without exception, and besides, it is often possible to work out probable derivations once certain common elements are known and some likely sources investigated in examples.

Map references such as Tavy Cleave (SX 5583) mostly refer to the Ordnance Survey Outdoor Leisure map 28 which covers the whole of the National Park using the scale 2½ inches to 1 mile (4cm to 1km). SX is the area grid reference.

The Author

Andrew Guy Stevens was born in Crewkerne, Somerset in 1945. He is the grandson of Philip Guy Stevens whose ink drawings illustrated William Crossing's famous *Guide to Dartmoor*. His early love of the moor was nurtured through annual family holidays which included marathon walks with Eric Hemery, author of *High Dartmoor*. He moved to Devon in 1979 and is currently Head of the Music Department at St Thomas' High School in Exeter.

◆

Common Elements

The River Dart which gives its name to the area is a Celtic word meaning Oak, and if you go to Wistman's Wood (SX 61 77) on the West Dart, or Dartmeet (SX 67 73), where the East and West Dart meet, the name makes obvious sense.

Short names, many of the rivers for example, have simple Celtic or Anglo-Saxon origins, but these and other ancient words are frequently combined to produce the place names that we find on modern maps. However, it must be said that change is a continuous process, and there are many place names on current Ordnance Survey maps which are different from the names in use ninety or a hundred years ago.

Obviously a knowledge of local social, religious and industrial history will help unravel the origins and meanings of place names: several authors have already done this most thoroughly (see page 32). And so, armed with some information and a little imagination, you can turn research back on itself so to speak and develop some insight into the history of the places you visit and the features you may observe, whether by choice or by chance.

One technique that might be useful if you are on your own or with 'understanding' friends, is to say the place names, with and without a Devon accent, and thereby appreciate how spoken sounds may change through useage and eventually lead to their being written differently. For example, the current word for boulder debris covering the slopes of a tor is CLITTER. Originally it was CLATTER. The local accent tends to soften the first vowel sound in this case.

BALL - often in the name of a hill which has a distinctly rounded shape.

BARTON - principal farm of a group in that it provided a central store, for barley in particular. Often known as the 'home farm' and frequently an enclosed design with a gated archway into the central yard.

BEACON - there are several hills on Dartmoor which incorporate this term. Some imply their earlier use as places where beacon fires were lit owing to their prominance, but others are probably derived from other local place names. For example, Ugborough Beacon (SX 66 59), might have been Peak Down Hill or Picken Hill originally.

BEAM - this always has some mining significance, either directly to deep, open workings where beam engines would have operated the mining equipment, or later on, to the adjacent hills.

BEARE or BERE - occasional association with the meaning 'fortified place' (Old English - *burh*), but more usually a reference to a wooded area of some sort.

BLACK - a very common component in place names, but not as obvious in meaning as might at first appear. The Anglo-Saxon word *blaec* means pale or colourless and both black and bleak are derived from this word. Bleak tends to imply the pale desolation of exposed hillsides whereas black implies the dark appearance of shaded valleys or woodland. Thus the two extremes of colourless can be interpreted.

BUCKLAND - common throughout Devon, this is a corruption of 'book land' and implies that the land was registered as being held by Royal Charter, dating from Saxon times.

COOMBE or COMBE or COMB - from the Celtic word *cym* which means a small valley usually closed at its upper end. The change to 'cum' or 'cam' is obvious but to 'ham' less so. Birkham Gate (SX 57 67) is not named on any map but it is also known as Burracombe Gate. With Burrator close-by, the whole area might have been known as Burra- or Bearacombe which simply means 'wooded valley'- which is what it is.

CLEAVE - as in Tavy Cleave (SX 55 83); in other words the valley through which the river Tavy flows, but since the word is probably derived from the Anglo-Saxon *cleof* meaning cliff, it describes the steep sides of the valley rather than the valley itself.

-DON - is simply a shortened version of 'down', but since 'don' and 'ton' sound almost identical as endings to longer place names, it is not surprising that the two versions have come to mean the some thing in some cases. For example Skerraton (SX 70 64) was *Sciredon* in records dating back to the 13th century. There are many examples of tautology in place-names, which have developed through time, and nearby Skerraton Down is just such a case. (Coombe Valley, just north of Bude in Cornwall, is a perfect example of this.)

FORD - implies a way of some sort, often across a river or stream, but not necessarily so. Sandy Ford, on the ancient burial path (Lich or Lynch Way) from Postbridge to the nearest Parish Church at Lydford, is devoid of river crossings. It is not marked on any maps, but the path descends through Sandy Ford to Hill Bridge (SX 53 80) from White Barrow (SX 56 79) between Langstone Moor and the newtake walls to the North. (see also WORTHY).

-HAM - one of several part-words indicating farm, homestead or nucleated village. This useage tends to be more common as you go south. Indeed the whole of the area to the south-east of Dartmoor is known as the South Hams.

LAKE - areas of standing water are always known as 'pools' on Dartmoor, whereas 'lake' refers to a small contributary stream. Dead Lake (SX 56 78) is a typical example. So is Red Lake (SX 64 66) although it should be noted that the 'pool' that is obvious in that area is the result of comparatively recent china clay workings.

-LEIGH (LEY) - this suffix refers to a woodland setting, or rather a natural clearing in woodland.

NEWTAKE - originally there were 35 ancient tenement farms that lay within the bounds of the Forest. There were certain rights attached to these tenements, including the right to enclose up to eight acres of land. These rights continued up to 1796, but since then a great deal of moorland, including common land, has been enclosed regardless of 'rights'. All such enclosures, old and new, are generally known as 'newtakes'.

POUND - usually a reference to a Bronze Age settlement where a roughly circular wall enclosed an area where both humans and their animals were at least partially protected. Therefore it is normal to find hut circles and other evidence of human habitation within the pound wall. DRIFT POUNDS were different in that they were enclosures constructed solely for the purpose of empounding cattle after they had been rounded up from the open moor.

STOCK - from the old English word *stoc* means 'secondary settlement', implying a farm other than the home or Barton Farm. Thus Tavistock would simply be 'farm on the River Tavy'. The fact that there was an abbey at Tavistock and that the old English word for 'holy place' was 'stow' might be confusing at first glance. The town is outside the National Park boundaries, but has great significance to Dartmoor in mining terms, since it was one of the four stannary towns where ore from the mines was brought to be weighed and stamped.

-TON - simply denotes 'farm' or 'enclosed area', but nothing in terms of size to the present-day derivative 'town'.

WELL - this word was sometimes used to describe natural springs, the only true well on the open moor being Fitz or Fitz's Well (SX 59 93).

WARREN - there are several place names incorporating this word; e.g. Hentor Warren (SX 59 65) and Ditsworthy Warren (SX 58 67). In each case this refers to the systematic management of rabbits for food, in comparatively recent times. The warrener would build large banks of stones and earth covered in turf to encourage the wild rabbit population to colonise. There are several ruins of the warreners' houses to be seen as additional evidence of these enterprises.

-WORTHY - from the Anglo-Saxon for 'enclosed homestead'. However there the simplicity ends because, of all the components of Dartmoor place names, this has evolved the most variants, including -ary, -ery, -over, -iver, -ever, even to -ford. North Hessary Tor above Princetown (SX 57 74), was more commonly known as Hisworthy Tor as recently as the beginning of the twentieth century.

Wheal Betsy, Mary Tavy. The engine house of what was once the most important lead and silver mine on Dartmoor, now owned by the National Trust

Alphabetical list of place names

ABBOTT'S WAY - (SX 68 65 and 61 67). There is an Abbey at Buckfast and there was one at Tavistock. Certainly the monks would have walked from one to the other, but whether they were responsible for the track in the first place is doubtful. The route is at least partially marked by crosses, although it is not certain that these were erected by the monks. SIWARD'S CROSS, better known as NUN'S CROSS (SX 60 69), is perhaps the best known of these, but there is evidence to suggest that this had been set up as a boundary rather than way marker and that 'Nun' is a corruption of its 17th century name *Nannecross*. *Nans* was a Celtic word for valley or dale.

Siward's (Nun's) Cross

ASHBURTON - (SX 75 69). Ash is a tree; burn is a stream; and ton is a homestead. From its origins as 'a small settlement by the stream where ash trees grow', it grew to become one of the principal towns serving the Dartmoor mining industry. Indeed it became one of the four stannary towns and there were important tracks to it from Plymouth and Tavistock, which were marked with guide-stones.

AVON - (SX 65 69). From the Celtic word meaning 'river'. Several place names off the moor incorporate its name as a locational reference.

BEARDOWN MAN - (SX 59 79). This is one of Dartmoor's tallest 'menhirs' or standing stones. 'Man' probably comes from the Celtic word *maen* meaning 'stone', but could simply be that such a stone looks like a figure, seen from a distance against the sky. Beardown ought to mean 'wooded hill', and Beardown

Tors are a little to the south and above Wistman's Wood which, at one time, was considerably more extensive.

BELLEVER - (SX 65 77). There is a tor, an ancient tenement farm and a well-known clapper bridge, each bearing this name. In early Duchy records the name was Welford which would seem to refer to the river crossing. There were several later corruptions of this; Wellaford (1579): Bellabur (1608); Bellaford (1663); and Bellefor (1736). William Crossing was using the 1663 spelling in his *Guide to Dartmoor* which was first published in 1909. I'm told that there is still a road-sign with this spelling in the area!

Clapper Bridge at Bellever, 1910

BELSTONE - (SX 62 93). If you believed the article in the *Western Morning News* during 1890, then you would be of the opinion that this name refers to a stone dedicated to the Phoenician god Baal! In Doomsday however, the place was recorded as *Bellestham. Belles* was a common name at that time and the terminal 'ham' can refer to an adjacent river bend, but more commomly simply means 'enclosure'. However, the English Place Name Society has the Doomsday spelling as *Bellestam* - a compound of two Old English words *belle* and *stan*, meaning the bell stone. This was supposed to make reference to '...a remarkably fine logan rock that rolled like a ship in a gale.' (*Little Guide Series*, Devon, 1900).

BITTAFORD - SX (66 57). The 'ford' is self-explanatory, but I can find no explanation of 'Bitta-'. It is not a reference to a river name, since the ford (and well-known bridge) is across the Lud Brook. It is probably another example of a personal name becoming part of a place name.

BOWERMAN'S NOSE - (SX 74 80). This massive granite pile of rocks, over thirty feet in height, has weathered in such a way that, from a distance, it can resemble a human profile. There are several stories about the origin of the name, but none has been properly substantiated. All that really matters is to make the distinction between this natural formation that suggests human characteristics, and the standing stones and crosses on Dartmoor which people have erected to serve some purpose or other.

BLACKINGSTONE ROCK - (SX 78 85). Probably this was originally just Blackstone, - which is self explanatory; thence via the pronunciation Blackystone to the current spelling. The 'Rock' part is superfluous. Legend has it that this and the other similar neighbouring rock piles were formed when King Arthur and the Evil One threw quoits at each other in battle, the missiles changing to rocks as they landed!

BRENT TOR - (SX 47 80). This is well outside the perimeter of the National Park, but it is a tor (with a church on top) and typifies the useage of the word 'brent' meaning 'prominent'.

BUCKFASTLEIGH - (SX 73 66). This is a compound of three Old English words; *buc* meaning 'deer', *feasten* meaning 'stronghold', and *leah* meaning 'clearing in the thicket'. Deer on Dartmoor were hunted to extinction by 1780, although they have recently recolonised some of the wooded valleys surrounding the moor, the Teign valley in particular.

BURRATOR WATERFALL - (SX 55 67). This is situated in the ancient woodland below the reservoir dam. The whole area was probably known as *Beara-*, *Bira-*, or *Burra-combe*, meaning 'wooded valley'. The -tor is really the high ground above. The more obvious waterfall (SX 55 68) is marked 'fall' on the map and is man-made. The Devonport Leat was originally constructed as a water supply for the Devonport area of Plymouth. When the original dam was opened in 1898, the lower section of the leat became redundant and the water was simply piped from its course along Yennadon Down and emptied into the reservoir.

CADOVER BRIDGE -(SX 55 64). The origins of this name go back to at least the 13th century. Cadworthy Farm has been so-called since this time, and the adjacent bridge was known as *Pontem de Cadaworth* in 1408. It is thought that the River Plym (hence Plymouth) was originally called the Cad. The river is easily fordable here, hence the suffix 'over' or 'way over the Cad'.

CATOR-(SX 68 76). Originally *Cadatrea* (1167), this is another place name that incorporates a person's name, in this case it is Cada's tree. According to a forester's account five hundred years ago, two of the five *villes* in the parish of Widecombe were the 'hamlet of North Catrowe' and the 'villat of Higher Catrowe'. The *venvilles* were the farms on the outskirts of the moor that, on payment of a small annual fee or *fin ville*, had commoners' rights on the adjacent moor, (usually of pasturage and turbary; i.e. grazing and peat cutting). Where a tree is part of a place name, it usually holds some special significance as a meeting place.

CHAGFORD -(SX 70 87). *Chag* is an Old English dialect word for gorse or broom and only found in place name compounds. With reference to '-ford', there are two ancient bridges over the River Teign at Chagford and stepping stones at Rushford Mill Farm just outside (SX 70 88). Chagford was another of the four principal stannary towns in the 14th century. (For a fuller explanation, see *Crossing's Guide to Dartmoor* pages 34-35.)

CHALLACOMBE - (SX 69 79). Also spelt Chalnecombe until recently and derived from the 14th century *Chelvecombe*, or 'valley of calves'. This was another of the venville farms. (see **CATOR**, and, for a fuller account, refer to *Crossing's Guide to Dartmoor*, pages 38-39.)

CHILDE'S TOMB - (SX 62 70). 'Cild' was a common Saxon appellation, and *Childe* a title of honour in the Middle Ages, but the story, or rather, legend associated with this Christian cross set beside a pagan kist and surrounded by a stone circle , was documented by Risdon nearly four hundred years ago, but is still popular myth. Childe the Hunter was caught in a blizzard and in a futile attempt to save himself, sacrificed his horse, disembowelled it and sheltered in the warm carcass! (see *Place Names of Plymouth, Dartmoor and the Tamar Valley* by W. Best Harris for a more comprehensive account.)

COSDON HILL - (SX 63 91). In the 1240 survey of Dartmoor known as the Perambulation, the name was *Cossdonne* - a compound of the personal name *Cos(s)* and *don(ne)* meaning 'hill'. Local pronunciation of the word is 'Cosson'. Between 1888 and the 1986, Ordnance Survey maps have incorrectly used the name 'Cawsand', but the recent Outdoor Leisure edition has reverted to the correct spelling.

CRANMEER POOL - (SX 60 85). One of the best known place names on Dartmoor, partly because of its bleak remoteness (even if approached by the military road from Okehampton), and more recently because it is the site of one of the original 'letter-boxes'. Indeed, its housing is really the only feature of the place since there's precious little pool any more. Until the recent craze for 'letter-boxing'- following clues and sampling the rubber stamps - the idea was to achieve the remote location, (Fur Tor -SX 58 83, and Duck's Pool -SX 62 67 are two other such places) and leave a stamped postcard in the letter-box for the next visitor to collect and post. Originally the name was *Crau Meer* meaning 'lake of the Crows'.

Saracen's Head Inn, 1797

DREWSTEIGNTON - (SX 73 90). Drew, as a family name can be traced back to 1210 in its original form *Drogo*. So the modern name is simply 'Drew's farm by the River Teign'. Nearby Castle Drogo became the most recent family home when it was completed in 1930. Until very recently, when the National Trust took over much of the estate, the family continued to own property in the village and in Broadhembury to the east (SX 10 04). Both villages have inns called the 'Drewe Arms'.

DRIZZLECOMBE - (SX 58 66). According to Crossing in 1909, Drizzle is a vernacular corruption of 'Thrushel'. Crossing refers to an old map in his possession which uses this name, although he recognises that the Ordnance Survey maps of the time were using 'Drizzle'. The location is important as one of the richest sites on Dartmoor for ancient relics, including fine stone rows and a great menhir (standing stone), all of which had become prostrate, but were re-erected in July 1893.

ERME - (SX 62 66). An interesting river name in as much as that the town Ermington probably gave its name to the river rather than the other way round. This, like the Plym, is known as 'back formation'. The Old English word *iermen* means 'main' or 'principal', so Ermington would have been the most important farm in the area (SX 63 53).

FERNWORTHY - (SX 66 84). It seems that 'fern' is a reference to bracken, which is the commonest type of fern and abundant on Dartmoor. Originally, Fernworthy was a group of three small farms with enclosures: now it is a reservoir surrounded by a large conifer plantation! In years of severe drought, it is possible to cross the South Teign by Fernworthy Bridge or the adjacent single stone clapper bridge, and thereby complete the ancient way from Metherall (SX 67 83) through Fernworthy to Froggymead (SX 65 84).

FINGLE BRIDGE - (SX 74 89). This narrow, three-arched structure is the old pack-horse way across the River Teign, and so-named because the Fingle brook enters the river immediately downstream. The modern spelling seems to date back to 1765, and the name had probably developed from the Old English word *fang* which meant to 'hold or catch'; which might suggest that the location was known for its fishing. Of course, it is cars which attempt the crossing now to the parking facilities beyond (most of them successfully), for the area has become a popular so-called 'beauty spot'. Incidentally, the 'V' shaped recesses which form part of the parapet, were designed to 'streamline' the structure in times of exceptional flood.

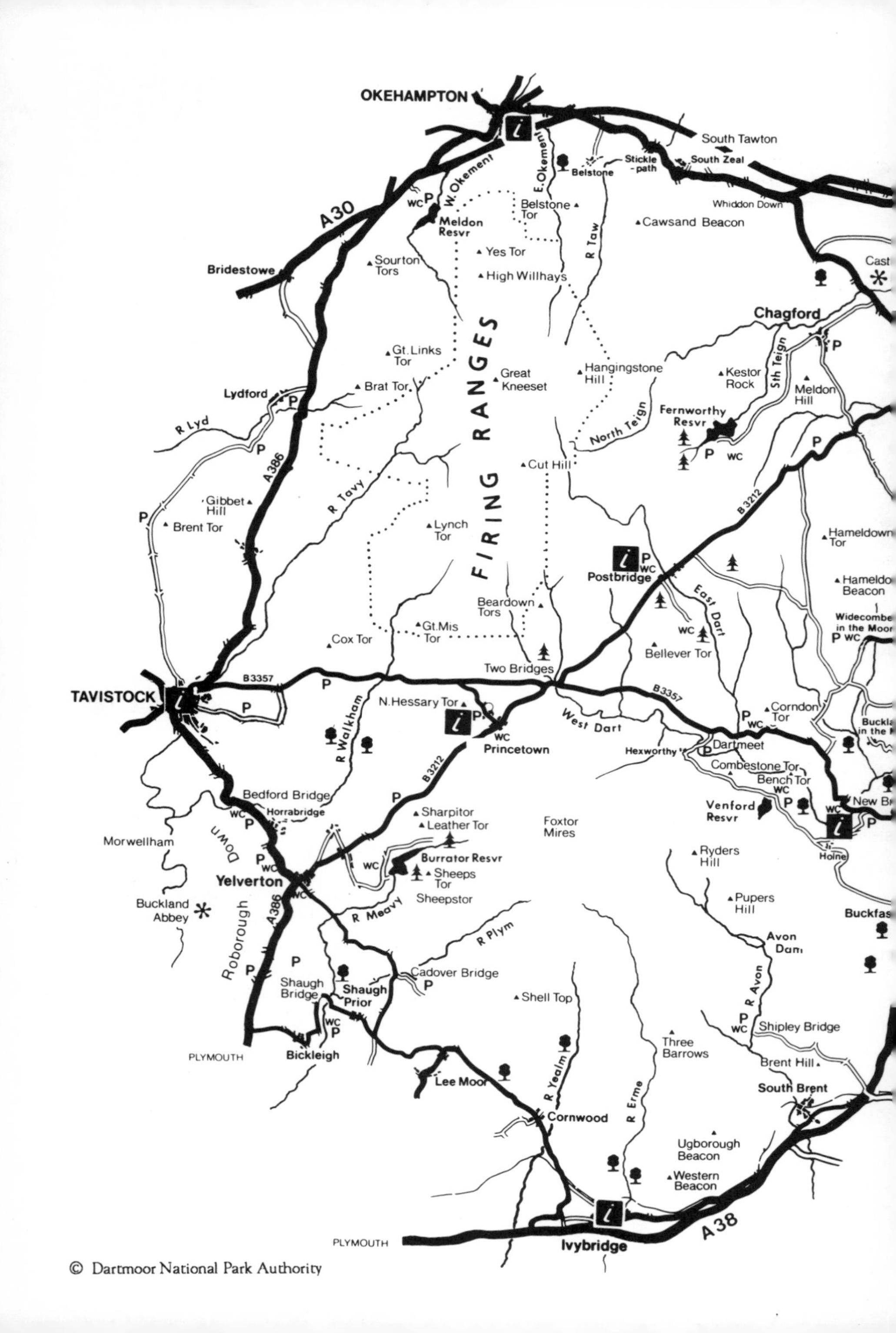
OKEHAMPTON
South Tawton
South Zeal
Stickle-path
Belstone
Whiddon Down
W. Okement
E. Okement
Belstone Tor
Cawsand Beacon
Meldon Resvr
A30
Yes Tor
R Taw
Sourton Tors
High Willhays
Bridestowe
Chagford
FIRING RANGES
Gt. Links Tor
Great Kneeset
Hangingstone Hill
Kestor Rock
Sth Teign
Meldon Hill
Brat Tor
Lydford
R Lyd
North Teign
Fernworthy Resvr
A386
Cut Hill
R Tavy
B3212
Gibbet Hill
Brent Tor
Hameldown Tor
Lynch Tor
Postbridge
East Dart
Beardown Tors
Gt. Mis Tor
Cox Tor
Two Bridges
Bellever Tor
TAVISTOCK
B3357
N. Hessary Tor
R Walkham
Princetown
West Dart
Corndon Tor
Hexworthy
Dartmeet
Combestone Tor
Bench Tor
Bedford Bridge
Horrabridge
Sharpitor
Leather Tor
Foxtor Mires
Venford Resvr
Holne
Morwellham
Down
Burrator Resvr
Sheeps Tor
Sheepstor
Ryders Hill
Yelverton
Buckland Abbey
Roborough
R Meavy
R Plym
Pupers Hill
Avon Dam
Cadover Bridge
Shaugh Bridge
Shaugh Prior
Shell Top
R Avon
Shipley Bridge
Three Barrows
Bickleigh
PLYMOUTH
Brent Hill
Lee Moor
R Yealm
R Erme
South Brent
Cornwood
Ugborough Beacon
Western Beacon
A38
Ivybridge

DARTMOOR

KEY

i Dartmoor National Park Information Centre

P Parking **WC** Toilets

EXETER
A30
R Teign
B 3212
Dunsford
Steps Bridge
P WC
Bridford
▲ Blackingstone Rock
onhampstead
Christow
Kennick Trenchford & Tottiford Resvrs
P
Hennock
Lustleigh
ecky Falls
EXETER
HQ
Bovey Tracey
Ilsington
A38
NEWTON ABBOT
NEWTON ABBOT
SHBURTON
TLEIGH

ES & TORBAY

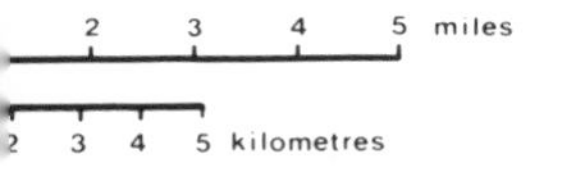

These days they perform a useful function as escape bays from the 'stream' of vehicles negotiating the bridge!

GIDLEIGH - (SX 67 88). Literally 'Gydda's clearing'. Although there is no person of this name on record, various spellings of the place name date back to the 12th century. William Crossing makes reference to the 14th century castle, and wonders at what period the Gidleys came into possession of the nearby manor.

GREY WEATHERS - (SX 70 80). Much research has been done on these two fine stone circles on the side of Sittaford Tor, before and after their partial restoration, but only Burnard in 1890 makes any reference to the meaning of their name. He says that they are "...thus named from their fancied resemblance to a flock of sheep."

GRIMSPOUND - (SX 70 80). This is probably the most visited and best preserved Bronze Age enclosure on the moor. There are no references to its name until the end of the 18th century, but 'Grim' in several other place name contexts is always a reference to the Devil, and suggests that the Saxons associated such a large prehistoric settlement with diabolic forces.

HARFORD - (SX 63 59). In 1086, the name was Hereford. The Old English word *here* meant 'army', and *har* must be a corruption of this. Unless the name goes back to some confrontation between the Celts and the invading Saxon forces, the 'army' reference cannot be explained. The bridge at Harford is over the River Erme.

HARFORD BRIDGE - (SX 50 76), is over the River Tavy above Tavistock and has a different derivation, being a shortening of the earlier spelling 'Hartforde', and being a possible reference to the spot being a place where deer would cross the river.

HEXWORTHY - (SX65 72). The '-worthy' as usual implies an enclosed home-stead, and the 'Hex' is a shortening of the personal name *Hexta* or *Hexte*; (15th century) - originally *Hexten*.

HOLNE - (SX 70 69).The name simply means 'place abounding in holly'. Certainly the area is wooded and the village feels sheltered. The fine church once belonged to Buckfast Abbey. The Church House Inn has a lengthy history as well, and there is a gravestone in the churchyard dedicated to Edward Collins, one time landlord of the inn, who died in December 1780. The inscription reads:

Here lies Poor Old Ned,
On his last Mattrass bed,
During life he was honest and free
He knew well the Chace
But has now run his Race
And his name was COLLINS D'ye fee.

HORRABRIDGE - (SX 51 69). Although Horrabridge is now a small town, it is the actual bridge over the River Walkham that is significant. 'Horra' is derived from the word *har*, meaning 'boundary', and the bridge is at the boundary of three parishes. Indeed there is a boundary stone on the bridge itself, set into the central parapet, facing downstream.

HUCCABY - (SX 66 72/3) Some of the early versions of this name perhaps provide a clue as to the original meaning: 1296 - *la Woghebye*; 1317 - *Woghby*: 1340 - *Woghebi*: 1417 - *Hogheby*: 1608 - *Hookeby*. The Old English word *woh* meant 'crooked' and *byge* meant 'bend ' or 'curve'. Certainly there is a big loop in the West Dart River here, separating Hexworthy from Huccaby on the other bank. However, Crossing points out that on the Hexworthy side of the river are some old farm enclosures know as the Byes,and notes that the '-by' in Huccaby is pronounced the same way by the locals; (rather than '-bee'). You have to draw your own conclusions in this case!

LETTAFORD - (SX 70 84). The Old English word *hluttor* meant 'clear, pure, bright', and so it would seem to indicate a crossing place of the stream at this point.

LUSTLEIGH - (SX 78 81). The English Place Name Society can only suggest that 'Lust-' is a derivative of the nickname *Luvesta* meaning 'dearest one', the '-leigh' as usual meaning 'clearing'. The community is outside the Dartmoor venville (see page 13, **CATOR** and *Crossing's Guide to Dartmoor* pages 38-39), but well within the National Park boundary and popular as a charming village in its own right and for its local scenery, in particular LUSTLEIGH CLEAVE which is the steep-sided ridge separating the Wray valley from the Bovey.

LYDFORD - (SX 51 84). Literally 'way over the river Lyd'. The river name has evolved from *aqua de Lide* (1249), through *Lidde* (1577) to its present form. It means 'noisy, loud stream' and certainly lives up to its name as it forces a way through Lydford Gorge. The Parish of Lydford is the largest in England, over 60,000 acres, including the whole of Dartmoor Forest, as fixed in the perambulation of 1240, and remaining practically unaltered to this day. Lydford was once the stronghold of King Alfred against the Danes, hence the castle, now in ruins. More recently, the village was served by two railway lines: the G.W.R single-track branch line from Plymouth, Marsh Mills to Launceston; and the S.R. double-track, Plymouth - Exeter - Waterloo main line. There are several road bridges left as evidence and a fine arched viaduct hidden away just to the east of the village (SX 514 846). The platforms of the combined station are overgrown but intact, and the site, undisturbed since 1968, has become a perfect and peaceful nature reserve.

MANATON - (SX 74 81). *Maene* was an Old English word meaning 'general' or 'common', and as part of a place name probably refered to an object that belonged to the whole community, or to a boundary marker that was common to adjacent estates.

MEAVY - (SX 54 67). The village simply takes its name from the river on which it stands. The name itself, also spelt Mewey, might well be pre-English, although the compound of Old English *maew* and *ie* would translate as 'Sea-gull river'. Close to the church gate is the famous ancient hollow oak tree, now supported and slowly dying; and underneath, the cross, which having been lost for over a century, was replaced on its pedestal (or calvary) and more recently restored. Don't be confused by the existence of a second large oak tree; that only goes back as far as 1918, the granite war memorial having been placed alongside a year later.

MELDON - (SX 55 92). The '-don' of Meldon is not a problem, but the 'Mel-' is!. The Old English adjective *maele* means 'spotted' or 'variagated': the old Celtic *mailo* means 'bare' or 'bald': and then *Maela* was also a personal name. Take your

Meavy, 1890, from Robert Burnard's *Dartmoor Pictorial Records*

Meavy, *c* 1954

pick! The English Place Name Society (*The Place Names of Devon* - Part 1, pages 203-204) discusses the derivation at some length without really coming to any firm conclusion. That apart, the actual hamlet of Meldon is tiny and the name these days is more usually used with reference to the huge quarry above the hamlet where vast quantities of stone are produced as track ballast for British Rail. The railhead for this operation is actually on the viaduct which crosses the Okement ravine, although no locomotives or loaded wagons may use the track. Further up the valley is the concrete dam of Dartmoor's newest and most controversial reservoir.

MERRIVALE - (SX 54 75). This name means 'pleasant, open space'. Indeed, there is precious little habitation here, apart from the Dartmoor Inn and the last working granite quarry on the moor. The name is also used to indicate one of the Army firing ranges in the area to the north, and the term 'Merrivale Antiquities' refers collectively to the abundant Bronze Age remains in evidence nearby. (SX 55 74).

MORETONHAMPSTEAD - (SX 75 86). In the Doomsday Book of 1086, the place is recorded as *Morton*, which is obviously 'moor farm'. The evolution to Moreton- is understandable, but there seems to be no explanation of the suffix 'hampstead', although the word is derived from the Saxon *ham-stede*, the place of the house. Indeed, the original two-syllable version is often used locally, and it would assist local road sign writers considerably if this version were revived officially!

OKEHAMPTON - (SX 58 95). This town is situated on the edge of the moor where main road and railway seem determined to squeeze a passage around the northern perimeter, at a point where the West and East Okement Rivers join. Of course the town takes its name from the rivers. In 1244 the river name was *aqua de Okem*, and *Ockment* by 1577. The first part of the name is from the Celtic *aku* meaning 'swift'. The second part may be derived from the Old Welsh *myned* - 'to go', or from the Aryan word *mim* meaning 'noisy'. Either way the picture of a fast flowing river is clear: indeed, both East and West Okements do descend rapidly from the high moor. It is interesting to note that the 1885 Ordnance Survey of Dartmoor uses the current spelling of the river name - OKEment - whereas writers of that period and for some time after were still using the old spelling - OCKment. The terminal '-hampton' is quite common and eventually replaced the Doomsday version *Ockmentune* or 'farm on the river Ockment'. Incidentally, 'Ockington', the local spoken corruption of the place name, has survived to this day.

PETER (MARY) TAVY - (SX 53 77 and 50 79). Although neither of these villages is actually adjacent to the River Tavy, they are close enough to have adopted the river name, combined with the names of the Saints to which their churches are dedicated. In each case the two elements should be pronounced as one word with the first element accented.

POSTBRIDGE - (SX 58 73). The 'clapper' bridge at Postbridge is probably the best known of its type on the moor, mainly because of its accessibility. The suffix 'Post' probably only dates back as far as the construction of the three-arched road bridge alongside. This new bridge would have been an important link in establishing the post-road from Moretonhampstead.These days, the garage and store close by the bridge also doubles as the local Post Office, and it was from there, right into the middle 1960's, that Mr Jack Bellamy would deliver the post on horseback - the last mounted postman in the country. Jack's son Reg, who has maintained his links with the village, recalls hearing the supporting piles of stones in the clapper bridge itself refered to as 'pawsts', and certainly this would be the pronounciation of 'post' with a broad Devon accent.

Postbridge, 1889

PRINCETOWN - (SX 58 73). This is a comparatively new settlement and developed around the notorious prison which was built between 1805 and 1809 to house French and American prisoners of war. A simple explanation of the name would be to accept that it was an association with the Prince Regent, afterwards George 1V. However, William Crossing, who, at the turn of the century, had several connections with the people of the town, did not accept such a straightforward explanation. PRINCE HALL (SX 62 74), now a Hotel, was formerly one of the ancient tenements mentioned in the earliest Forest records. At about the same time that Sir Thomas Tyrwhitt was building a large residence for himself at TOR ROYAL (SX 69 78), a Judge Buller had acquired Prince Hall for development. He was also responsible for extending and developing TWO BRIDGES (SX 60 75), adding some cottages and the Saracen's Head Inn. When Arthur Young, the authority on agriculture, visited the area in 1796 he made reference to Two Bridges as Princetown. Quite apart from this, the prison site was always known as Dartmoor Prison and this is confirmed, for example, in a letter written by an official of the Transport Board in 1805, in which there is no reference to the prison site being at Princetown. Thus Crossing was suggesting that the name had been transferred from the developing community at Two Bridges. Either way, the prison town did acquire its current name, whatever its origin, and this is confirmed by an indisputable reference in the baptismal register (1812):-

'October 4th Ann, daughter of William Robins and Ann his wife, resident at Prince Town Brewery.'

After this date, the place name often appears in this register, although it seems that it was not more generally used for some time.

Prince Hall, 1797, from an 18th century water colour

RUNDLESTONE - (SX 57 74). According to William Crossing, this small collection of buildings, high up on the road between Two Bridges and Tavistock, takes its name from the granite pillar, the Rundle Stone, that once marked the Royal Forest boundary. It had been recognised as such in 1702, although it had not been mentioned in surveys of the area prior to this date. The stone stood on the south side of the road immediately opposite the more recent Lydford/Walkhampton parish boundary-stone. Crossing measured the stone in 1881:

> *'It stood 7 feet above the stones in which it was set, and was four feet in girth. Near the top was the letter R cut in relief. It is marked on a map dated 1720 as a "Great Stone call'd Roundle."'*

(The stone was broken up some years after 1881 to provide building material for a nearby wall.)

Hansford Worth, however, writing some 40 years later, challenges Crossing's explanation. In particular, he notes that, apart from Nun's Cross (SX 60 69), all the Forest Boundary markers were natural objects and that the so-called Rundle Stone was not strictly on the line of the boundary anyway. Worth takes up the 'roundle' version of the word. In everyday parlance and in heraldry, it means 'small circular object'. He suggests that nearby Rundlestone Tor (SX 57 74) which is on the Forest Boundary line, is the real Roundle Stone - especially because of its unique rock-basins. One of these 'roundles' was formed before the whole main slab moved from a level position, allowing a second basin to form near the summit on the new level surface. He concludes - "This rock, with its considerable area, its thickness of 4' 6", and the circles of its rock basins, certainly agrees with the description given in 1736 - 'a Great Stone call'd Roundle'." (Until the most recent revision of the Ordnance Survey maps, the place name had mysteriously been spelt Rendlestone for several decades.)

SAMPFORD SPINEY - (SX 53 72). This tiny village is just to the west of the river Walkham, the first part of the place name being a reference to the river crossing, a sandy ford, 'Samp-' being a corruption of the 1086 spelling *Sanford(a)*. Spiney refers to Gerard de Spineto who held the manor in the early 13th century.

SHAUGH PRIOR - (SX 54 63). Shaugh is a typical Dartmoor border settlement, with the buildings grouped around the solid old parish church. There have been several spellings and pronounciations of the name, all of which relate to the Old

English word *sceaga*, which means 'wooded', a description which certainly still applies today. The manor was from the earliest times owned by Plympton Priory and this would account for the second part of the name.

SHEEPSTOR - (SX 55 67). Although the connection between the village and the tor close by is obvious, neither has anything to do with sheep - except possibly their droppings! The name of the tor, however, seems to have evolved separately, since it was established as *Shitestorr* by 1291. There were many recorded versions of the village (and parish) name - including *Shistor* (1547); *Shepystor* (1574); and *Shittistor* (1691) with its variant *Shitteslowe*. It has been suggested that all these originate in the Old English *scyttel*, meaning 'bar or bolt' with the suggestion that the tor resembles such an object when viewed from certain vantage points. It has also been suggested that the change from a 't' sound to a 'p' sound might have been a deliberate move away from any lavatorial connotations!

SHILSTONE - (SX 70 90). Shilston(e) itself is an ancient farming community, but the name is a reference to the famous dolmen or cromlech which stands in an adjacent field to the north. The name is compounded from the two Old English words *scylf* - meaning 'shelf', and *stan* meaning 'stone'. Together they define the cromlech with its three granite supports and huge, flat capping stone. There are five other similar place names in Devon, including the Shilstone at Throwleigh (SX 65 90), but no similar remains have been discovered; indeed the surviving one collapsed in 1862 and was restored later the same year. (Maps and signposts refer to the relic as 'Spinsters' Rock(s)', with reference to the legend that three spinster yarn-spinners had raised the monument one morning before breakfast! Of course there must be some doubt as to the type of yarn they used to spin!)

SOUTH BRENT - (SX 69 60). The word Brent is possibly an obscure derivative of the Old English word *brant* meaning 'steep', which would be a reference in this case to the prominent hill of the same name just to the north. The 'South' suggests that there was another settlement of the same name further up the river Avon to the north-west.

STICKLEPATH - (SX 64 94).This ancient village lies tucked in below the great mound of Cosdon Hill. There are several ways up to the moor from here, all of them steep climbs - and that's exactly what 'stickle' means.The original Anglo-Saxon word was *sticol* or *sticele*. The place name was recorded as *Stikelepethe* as early as 1280. The STICKLEPATH in the Walkham valley on the south west edge of the National Park (SX 49 70) is no more than a steep track leading down to Grenofen Bridge from Buckland Monachorum.

THROWLEIGH - (SX 66 90). Situated on the edge of the moor, *Throulegh* was one of the venville 'towns', in other words, a farm that had rights on the moor and surrounding commons. Although the modern spelling is different, the pronounciation is unaltered. The first part of the name is probably derived from the Old English *þruh*, meaning 'chest' or 'coffin', and may refer to the several cains and tumuli (ancient burial mounds) within the parish boundaries. The suffix 'leigh' as usual indicates a clearing.

WALKHAMPTON - (SX 53 69). Certainly this village is situated on the river Walkham, but there has been considerable conjecture amongst writers and historians as to whether the river name was taken from the place name, or vice versa. In his *Guide to Dartmoor* page 126, William Crossing summarises the controversy after noting that part of Walkhampton Common and part of the Chase of Okehampton were both held by the De Redvers family in the 13th century. In this case it would seem appropriate to quote Crossing in full.

> *'It is not a little curious that in the names of these commons the termination "hampton" appears, and yet is found nowhere else on the moor, or in the border parishes. Many of the names of the latter exhibit the oft-found Saxon termination "ton", though in more than one instance the word is apparently traceable to the Celtic "dun", a hill, the heavy sound of the initial letter having given place to a lighter one. But "hampton" is only found in Walkhampton and Okehampton - locally pronounced Wackington and Ockington - though in neither does the word seem to possess its usual signification.*
> *It would, however, be unsafe to conclude that it does not. "Ham" and "ton" may together, be taken to mean a farm, or enclosed land, with its dwelling-house and outbuildings, the "house town", as it were, and the term came to signify an inhabited settlement. In Walkhampton, the second syllable does not appear to have any connection with the third, but only with the first, the name being derived, we may*

reasonably suppose, from the river Walkham. Risdon, writing early in the seventeenth century, calls the river Store, but even if it were then so known, it is certain that at a much earlier time it bore a name closely resembling the one by which it is called to-day, being referred to as the Walkamp in the deed of Isabella de Fortibus. Thus, Walkhampton would mean the town, or settlement, on the Walkham, if we could be sure that the deed gave us the earliest form of the name of the river. But this is doubtful. There are many Dartmoor streams bearing the name of Walla, or Wella, and one that of Wollake, and I should be inclined to place Walkham in the same category, and to regard its early name to have been either Walla or Wollake. In Saxon times the settlement on the stream would be called Wallahampton or Wollakhampton, and by an easy transition Walkhampton. But it is also very probable that we do not see the word "ham" in this name at all; that the early name of the river was the Walla, and that Walkhampton is "Walla cwm ton", the town in the combe, or valley, of the Walla.'

WALLABROOK - There are several streams on Dartmoor with this name or known by this name. Whereas the second part, 'brook', has hardly altered in spelling, pronounciation or meaning from the Old English *broc*, there are several possibilities as to the origins of 'Walla'. The Anglo-Saxon *waela*, or *wielle* meaning 'well' would do, except that the various streams do not all emerge from distinct springs, which is what the word implies. Another possibility is that Walla is a corruption of *walter* meaning 'tumbling' or 'rolling'. Then there's the Anglo-Saxon word *wileg*, meaning 'willow', and that too could have been the origin.

The Wallabrook (SX 63 87) which is a tributary of the Teign was called *Wotesbrokelakesfote* in 1240. If *-oke-* is oak, and *-lake-* is stream, and *-fote* is foot, then this would refer to the last part of the stream before it joins the main river, where there are oak trees growing alongside.

The Western Wallabrook (SX 66 67), which is a tributary of the Avon, is spelt Wellabrook on the Ordnance Survey maps. This variant would be consistent with the notion that it is a stream distinctly emerging from a spring.

The Hentor Brook (SX 58 65) is sometimes called the Wall or Walla Brook, but this name is quite separate, since it is a reference to Willings Walls Warren which is immediately to the south-west.

WARREN HOUSE INN - (SX 67 80) There was no building on this site before 1720, and the present building, erected in 1845, replaced one on the other side of the Moreton to Postbridge road. This was known as Newhouse. The Warren part of the name is a result of its proximity to Headland Warren (see Common Elements). In fact Headland Warren House was at one time an inn (Birch Tor Inn) which was, no doubt, less remote in the days when there was extensive mining in the vicinity! Birch Tor (SX 68 80) was also known as Warren Tor. Leading down from this Tor to Bennet's Cross are a series of bond-stones marked WB for warren boundary. Indeed, the cross itself bears the same inscription, although its history as a route marker and parish boundary stone goes back considerably further. The present Warren House is still comparatively remote - it still has to generate its own electricity - and serves as a welcome rendezvous for the traveller, arriving on foot, horseback or by car.

WIDECOMBE-IN-THE-MOOR - (SX 71 60). All early versions of this name, going back as far as 1270, suggest the name to mean *withy* or 'wide' valley - a suitable name based on obvious physical location. This village, famous for its annual fair, and dominated by ninety feet of church tower, is set in the impressive East Webburn River valley, between Hamel Down to the north-west and Top Tor to the south-east.

Warren House Inn, 1913

WISTMAN'S WOOD - (SX 61 77). Descriptions of this small area of knarled oak trees emerging through moss-covered boulders suggest that it has changed little since the Norman conquest and the subsequent survey of the area - the so-called 'perambulation'. The name is perhaps a reference to the Devil - *wisht* being Devon dialect for 'eerie' or 'uncanny'. Certainly it is a lonely position, up the east side of the West Dart river from Two Bridges, and hugging the side of the valley below Littaford Tors.

Other derivations, however, have been suggested. The Celtic words *visg maen coed* which mean 'stony wood by the water', would make sense phonetically and in meaning. Until quite recently, the old locals spoke of the oak groves as 'Welshman's Wood'. Now this could have been a corruption of *Wealasmans Wood: wealas* meant 'foreigner', and so it could have been the 'Wood of the Celts' who were regarded as foreigners by the Saxon settlers.

The only other similar surviving oak wood is Black-a-Tor Copse(SX 56 89). This is situated on the north-east bank of the West Okement river, in an even more impressive and remote valley bottom - so remote it would seem that even the Ordnance Survey wasn't sure where it was, since, for many years the maps showed the copse as being on the opposite side of the river! (This mistake has more recently been rectified.)

YELVERTON - (SX 52 67). On the east side of Burrator Reservoir (SX 55 68) there is a peninsula on which are the extensive remains of what was obviously an important estate. This was Longstone and the seat of the Elford family from the end of the 15th century to the mid-18th century. The family also possessed property close by Roborough Down and this was called Elford Town. Local dialect gradually changed the pronounciation to Yelver Town or Yelverton, and it was this version that the Great Western Railway Company adopted for the station there, when the branch line from Marsh Mills (Plymouth) to Tavistock was opened in 1859. This is still the name that now describes the main part of the conurbation, but Elfordtown and the nearby farm of the same name, have survived as a separate entity.

Tavistock Turnpike, Two Bridges (SX 60 74)

Books

Baring-Gould, Sabine A *Book of Dartmoor* (Wild Wood House, re-issued 1982 Camelot Press)

Brunsden, D *Dartmoor* (Geographical Association 1968)

Burnard, Robert *Dartmoor Pictorial Records* (Brendon & Son 1890; Devon Books Facsimile Subscriber Edition 1986)

Cocks, John Somers A *Dartmoor Century* (One Hundred Years of the Dartmoor Preservation Association 1883 - 1983) (DPA 1983)

Crossing, William *Princetown - It's Rise and Progress* (Quay Publications 1989)
- *Guide to Dartmoor* (Facsimile of the 1912 edition Peninsula Press 1990)

Dartmoor National Park Authority *Dartmoor Place Names Index* (DNPA 1989)

Gover, JEB: Mawer, A: Stenton, FM *The Place Names of Devon, Parts I and II* (English Place Name Society Vol.VIII Cambridge University Press 1931)

Harris, W.Best *Place Names of Plymouth, Dartmoor and the Tamar Valley* (Stannary Press 1987)

Hemery, Eric *High Dartmoor* (Robert Hale 1982)
-*Walking the Dartmoor Railroads* (Peninsula Press 1991)
- *Walking the Dartmoor Waterways* (Peninsula Press 1991)

Page, John Lloyd Warden *Dartmoor and its Antiquities* (Seeley and Co. 1889)

St Leger-Gordon D *Under Dartmoor Hills* (Robert Hale 1954)

Worth, Hansford *Dartmoor* (Latimer,Trend and Co. 1953)